Negative

By Robert

An Easy Steps Math book

Copyright © 2014 Robert Watchman

All rights reserved.

No portion of this publication may be reproduced, transmitted or broadcast in whole or in part or in any way without the written permission of the author.

Other books in the Easy Steps Math series

Fractions
Decimals
Percentages
Ratios
Negative Numbers
Algebra
Master Collection 1 – Fractions, Decimals and Percentages
Master Collection 2 – Fractions, Decimals and Ratios
Master Collection 3 – Fractions, Percentages and Ratios
Master Collection 4 – Decimals, Percentages and Ratios

More to Follow

Contents

Introduction	7
Chapter 1 **Negative Number Basics**	9
Chapter 2 **Comparing Positive and Negative Numbers**	12
Chapter 3 **Adding Negative Numbers**	15
Chapter 4 **Subtracting Negative Numbers**	18
Chapter 5 **Mixed Addition and Subtraction of Negative Numbers**	21
Chapter 6 **Multiplication of Negative Numbers**	23
Chapter 7 **Division of Negative Numbers**	27
Chapter 8 **Combined Operations**	31
Multiplication Tables	33
Answers	35
Glossary of Useful Terms	37

Introduction

This series of books has been written for the purpose of simplifying mathematical concepts that many students (and parents) find difficult. The explanations in many textbooks and on the Internet are often confusing and bogged down with terminology. This book has been written in a step-by-step 'verbal' style, meaning, the instructions are what would be said to students in class to explain the concepts in an easy to understand way.

Students are taught how to do their work in class, but when they get home, many do not necessarily recall how to answer the questions they learned about earlier that day. All they see are numbers in their books with no easy-to-follow explanation of what to do. This is a very common problem, especially when new concepts are being taught.

For over twenty years I have been writing math notes on the board for students to copy into a note book (separate from their work book), so when they go home they will still know how the questions are supposed to be answered. The excuse of not understanding or forgetting how to do the work is becoming a thing of the past. Many students have commented that when they read over these notes, either for completing homework or studying for a test or exam, they hear my voice going through the explanations again.

Once students start seeing success, they start to enjoy math rather than dread it. Students have found much success in using the notes from class to aid them in their study. In fact students from other classes have been seen using photocopies of the notes given in my classes. In one instance a parent found my math notes so easy to follow that he copied them to use in teaching his students in his school.

You will find this step-by-step method of learning easier to follow than traditional styles of explanation. With questions included throughout, you will gain practice along with a newfound understanding of how to complete your calculations. Answers are included at the end.

Chapter 1

Negative Number Basics

Negative numbers, also known as directed numbers are positive and negative numbers. Knowledge of negative numbers is essential in everyday life and in mathematics as there is hardly a time when they are not being used. Temperatures go down below zero; people have debts, and businesses have debts and these are all values below zero.

All numbers are either positive or negative, except for zero, which is neither. Look at the number line below. All the numbers to the left of the zero are negative numbers. All the numbers to the right of the zero are positive numbers. The zero is neither positive nor negative. The arrows at both ends of the number line means that the number line continues on in both directions to infinity.

The numbers to the right are increasing in value and the numbers to the left are decreasing in value. So the number -10 is smaller than -5 and the number 3 is larger than -3.

You will notice that the positive numbers don't have a plus (+) sign but the negative numbers all have a minus (–) sign. This is because if there is no sign to the left of a number, then everyone will know it is positive.

Note that the + or – sign is always written to the left of the number.

Positive and negative whole numbers including zero are called **integers**. (Refer to the Glossary at the end for more definitions).

Remember; throughout this book the rules for PEMDAS or BODMAS applies.

Calculators do not always give the correct answer when solving these types of questions. This is because many calculators need to have the keys pressed in a specific order otherwise errors will occur, especially for negative numbers. Also it can take a long time to work out how to use the calculator for simple questions and using the calculator can sometimes take longer than just answering the question in your head.

In mathematics, three dots set out like ∴ is an abbreviation for therefore. This will be used throughout this book where appropriate.

Indicate whether the following are Positive (P) or Negative (N).

a) Making a profit in business.

b) Temperature going down.

c) Elevator going into the basement.

d) Depositing money into your account.

e) Owing money to the bank.

f) Losing money at the casino.

g) Getting a pay rise at work.

h) Withdrawing money from your account.

i) Temperature going up.

j) Going below sea level in a submarine.

Indicate the directed number for each of the following.

a) Making a loss of $550 in business

b) Diving 15 m underwater.

c) Winning a game by 7 points.

d) Depositing $400 into your bank account.

e) Temperature dropping 3 degrees.

f) A pay rise of $50.

g) Taking the elevator up 5 floors.

h) Making a profit of $1000 in business.

i) Losing a game by 3 points.

j) Jumping 3 feet.

Chapter 2

Comparing Positive and Negative Numbers

It is important to understand the values of positive and negative numbers and which numbers are the larger ones and which are the smaller ones. The greater than symbol (>) and the less than symbol (<) can be used to compare negative numbers and to show which is larger or smaller. Being able to arrange numbers in ascending (increasing) order and/or descending (decreasing) order also shows an understanding of this concept.

A number line is a great tool to use and helps greatly when comparing or arranging positive and negative numbers. It allows you to see where the numbers are when comparing or arranging numbers.

For example:

Saying −5 is greater than −10 is the same as writing −5 > −10 in mathematical terms. On a number line, the numbers to the right are always larger than the numbers to the left so you can see that since −5 is to the right of −10 then it must be greater. It is important to use a number line until you get used to how positive and negative numbers work. After that the number line will not be necessary.

Insert the correct symbol between the following pairs of numbers.

a) +9   +5

b) +3   −5

c) −12   −8

d) −7   +2

e) −3   −7
f) −12   −15

g) +9   −9

h) −3   +3

i) −1   −100

j) −3   7

The plus (+) sign has been left out of the questions below.

Arrange the following from smallest to largest

a) −3, −12, 6, −4, −1

b) −1, 1, −4, −2, −11

c) 49, −33, −78, −90, 2

d) 2, −8, −3, −9, −5

e) 76, −89, −7, 0, −12

Arrange the following from largest to smallest

a) −8, 7, 10, 1, −12

b) −2, −4, −9, −6, −7

c) −11, −5, −2, −12, −9

d) 60, −55, −81, 6, 0

e) 9, −48, 98, −99, −8

Chapter 3

Adding Negative Numbers

When you're first learning to add and subtract positive and negative numbers, using a number line is very important. Just draw a number line on the top of your page and use it when answering questions. It will allow you to see which direction you need to go when adding or subtracting.

There are two special rules to remember when you are adding positive and negative numbers, so you will need to memorise the following.

Whenever you see two plus (+) signs next to each other, the operation is addition.

For example:

Positive nine plus positive seven $+9 + (+7)$ [you are adding a positive number]. This is the same as $9 + 7$ which of course equals 16.

Step 1. Decide if the operation will be a plus (+) or a minus (−). Plus in this case.
Step 2. **Start with the first number on the number line (this is a 9),**
Step 3. Move to the right 7 places. This adds the 7 and you land on 16, which is your answer.

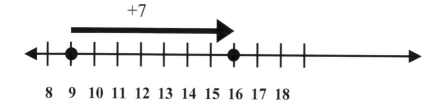

Whenever you see a plus (+) sign and a minus (−) sign next to each other, the operation becomes a subtraction.

For example:

Positive nine plus negative seven +9 + (−7) [you are adding a negative number]. This is the same as 9 − 7 which of course equals 2.

Step 1. Decide if the operation will be a plus (+) or a minus (−). Minus in this case.

Step 2. Start with the first number on the number line (this is a 9),

Step 3. Move to the left 7 places. This subtracts the 7 and you land on 2, which is your answer.

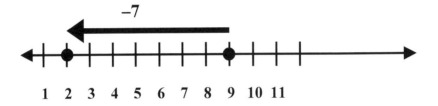

So when you add a positive number you are just adding and you move to the right on the number line. When you add a negative number, you are subtracting so you move to the left on the number line.

Note that you don't always need to draw a large number line, only draw the section you need. If the numbers get too large, it is best to visualise or see the number line in your head.

Add these Numbers

a) +10 + (+7)

b) +7 + (−5)

c) −3 + (+5)

d) +8 + (−10)

e) −1 + (−4) + (+5)

f) −10 + (−24)

g) −10 + (−2) + (+15)

h) +74 + (−50)

i) +34 + (−12) + (−8)

j) −29 + (+5) + (+6)

Chapter 4

Subtracting Negative Numbers

There are also two special rules to remember when you are subtracting positive and negative numbers, so you will need to memorise the following. These are similar to the rules for addition.

Whenever you see a minus (−) and a plus (+) sign next to each other, the operation is a subtraction.

For example:

Positive eleven minus positive eight $+11 - (+8)$ [you are subtracting a positive number]. This is the same as $11 - 8$ which of course equals 3.

Step 1. Decide if the operation will be a plus (+) or a minus (−). Minus in this case.

Step 2. Start with the first number on the number line (this is an 11)

Step 3. Move to the <u>left</u> 8 places. This subtracts the 8 and you land on 3, which is your answer.

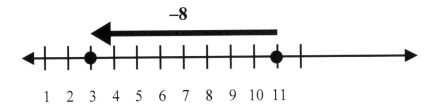

Whenever you see two minus (−) signs next to each other, the operation is addition.

For example:

Positive eleven minus negative eight +11 − (−8) [you are subtracting a negative number]. This is the same as 11 + 8, which of course equals 19.

Step 1. Decide if the operation will be a plus (+) or a minus (−). Plus in this case.

Step 2. Start with the first number on the number line (this is an 11)

Step 3. Move to the <u>right</u> 8 places. This adds the 8 and you land on 19, which is your answer.

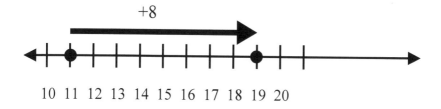

10 11 12 13 14 15 16 17 18 19 20

So when you <u>subtract a negative number</u> you are <u>adding</u> and you move to the right on the number line. When you subtract <u>a positive number</u>, you are <u>subtracting</u> so you move to the left on the number line.

It doesn't matter if the starting numbers are positive or negative. This just tells you where to start on the number line.

An easy way to remember the rules above is:

if the signs are the same, you add
+ (+) = + or **− (−) = +**

If the signs are different you subtract
+ (−) = − or **− (+) = −**

Subtract these Numbers

a) $+7 - (+8)$

b) $-3 - (-5)$

c) $+4 - (+7)$

d) $+8 - (-12)$

e) $-7 - (-2) - (+4)$

f) $+67 - (-13)$

g) $+17 - (+9) - (+10)$

h) $-22 - (-12) - (-10)$

i) $-75 - (+15)$

j) $+51 - (+43) - (-7)$

Chapter 5

Mixed Addition and Subtraction of Negative Numbers

Questions are usually not all addition or all subtraction. They are mixed and the numbers usually don't have parentheses around them, so you will need to be able to work without these.

For example, the following sets of questions are the same. The second set is how it looks without parentheses and without the positive and/or negative sign.

$+5 - (+3)$ is the same as $5 - 3$ which equals 2
$-6 + (-8)$ is the same as $-6 - 8$ which equals -14
$-12 + (+6)$ is the same as $-12 + 6$ which equals -6
$+2 + (+4) - (-5)$ is the same as $2 + 4 + 5$ which equals 11

Rewrite the following and fill in the blanks

a) $+12 + (+3)$ is the same as _____ which equals ___

b) $-5 - (-8)$ is the same as _____ which equals ___

c) $-2 + (-7)$ is the same as _____ which equals ___

d) $+8 + (+40)$ is the same as _____ which equals ___

e) $-87 + (+32)$ is the same as _____ which equals ___

f) $+4 + (+5) + (+6)$ is the same as _____ which equals ___

g) $-8 - (-7) + (-3)$ is the same as _____ which equals ___

h) $-6 + (-3) + (-2)$ is the same as _____ which equals ___

i) +2 + (−8) + (+3) is the same as _____ which equals____

j) +56 + (−11) − (− 4) is the same as _____ which equals ____

The following questions are the same as what you have done so far. The only difference is that the parentheses have been removed and there are be as many positive and negative signs.

Complete the following mixed questions

a) 3 + 6 + 2 =

b) −12 + 4 +3 =

c) 7 + 5 −6 =

d) −42 + 25 −9 =

e) −6 −3 −7 =

f) 9 −10 +7 =

g) −4 −9 +4 =

h) −22 + 39 −14 =

i) −5 + 7 − 3 =

j) 8 −10 + 9 =

Chapter 6

Multiplication of Negative Numbers

Just like for addition and subtraction, there are rules to follow for multiplying negative numbers.

When multiplying two numbers with the same sign the answer will be positive.

Therefore: $+ \times + = +$ and $- \times - = +$

When multiplying two numbers with different signs the answer will be negative.

Therefore: $+ \times - = -$ and $- \times + = -$

For example

Numbers with the same sign
+5 × +7 = +35 and −5 × −7 = +35

Numbers with different signs
+5 × −7 = −35 and −5 × +7 = −35

Don't forget you don't need to put the (+) signs in, but you must put the (−) signs in. So the multiplications above will look like this.

Numbers with the same sign
5 × 7 = 35 and −5 × −7 = −35

Numbers with different signs
5 × −7 = −35 and −5 × 7 = −35

Multiplying positive and negative numbers is a basic 3-step process.

Step 1. Work out what sign the answer will be

Step 2. Work out the product of the numbers

Step 3. Put the two together.

E.g. 1. Calculate 2×-5

Step 1. Work out the sign.

$+ \times - = -$, therefore negative $(-)$

Step 2. Work out the product of the numbers.

$2 \times 5 = 10$

Step 3. Put the two together. Answer is -10.

$\therefore 2 \times -5 = -10$

E.g. 2. Calculate -3×-7

Step 1. Work out the sign.

$- \times - = +$, therefore positive $(+)$

Step 2. Work out the product of the numbers.

$3 \times 7 = 21$

Step 3. Put the two together. Answer is $+21$.

$\therefore -3 \times -7 = 21$

If there are more than 2 numbers to multiply, just work from left to right. The sign may change more than once.

E.g. 3. $-3 \times -2 \times -1$

Step 1. Work out the sign. Work on the first 2 signs from left to right

$- \times -$ This gives a $+$.

Now multiply this new sign with the third sign.

$+ \times -$ This gives a $-$

So our answer will be a negative $(-)$

Step 2. Work out the product of the numbers.

$3 \times 2 \times 1 = 6$

Step 3. Put the two together. Answer is -6

$\therefore -3 \times -2 \times -1 = -6$

Calculate the following using the steps above

a) $-12 \times 7 =$

b) $3 \times -5 =$

c) $-10 \times -3 =$

d) $-5 \times -5 =$

e) $-8 \times 2 =$

f) –6 × 3 × –2 =

g) –3 × –4 × –5 =

h) –8 × –4 × 2 =

i) 2 × 5 × –5 =

j) 6 × 9 × –1 =

Chapter 7

Division of Negative Numbers

When dividing positive and negative numbers, the same rules apply as for multiplication.

When dividing two numbers with the same sign the answer will be positive.

Therefore: $+ \div + = +$ and $- \div - = +$

When dividing two numbers with different signs the answer will be negative.

Therefore: $+ \div - = -$ and $- \div + = -$

These rules apply to fractions also since fractions are a division. (Refer to the Easy Steps Math Fractions book)

For example

Numbers with the same sign
$+56 \div +7 = +8$ and $-56 \div -7 = +8$

Numbers with different signs
$+56 \div -7 = -8$ and $-56 \div +7 = -8$

Don't forget you don't need to put the (+) signs in, but you must put the (−) signs in. So the divisions above will look like this.

Numbers with the same sign
$56 \div 7 = 8$ and $-56 \div -7 = -8$

Numbers with different signs
$56 \div -7 = -8$ and $-56 \div 7 = -8$

Dividing positive and negative numbers is also a basic 3-step process.

Step 1. Work out what sign the answer will be

Step 2. Work out the quotient of the numbers

Step 3. Put the two together.

E.g. 1. Calculate $+20 \div -4$

Step 1. Work out the sign.

$+ \div - = -$, therefore negative $(-)$

Step 2. Work out the quotient of the numbers.

$20 \div 4 = 5$

Step 3. Put the two together. Answer is -5.

$\therefore 20 \div -4 = -5$

E.g. 2. Calculate $-35 \div -7$

Step 1. Work out the sign.

$- \div - = +$, therefore positive $(+)$

Step 2. Work out the quotient of the numbers.

$35 \div 7 = 5$

Step 3. Put the two together. Answer is $+5$.

$\therefore -35 \div -7 = 5$

If the question is a fraction, follow the same steps.

E.g. 3. $\dfrac{25}{-5}$

Step 1. Work out the sign.

$\dfrac{+}{-} = + \div - = -$, therefore negative (–)

Step 2. Work out the quotient of the numbers.

$25 \div 5 = 5$

Step 3. Put the two together. Answer is –5

$\therefore \dfrac{25}{-5} = -5$

Calculate the following using the steps above.

a) $-12 \div -3$

b) $6 \div -2$

c) $-44 \div -11$

d) $-49 \div 7$

e) $36 \div -6$

f) $-56 \div -8$

g) $\dfrac{48}{-3}$

h) $\dfrac{-45}{-5}$

i) $\dfrac{-84}{21}$

j) $\dfrac{54}{9}$

Chapter 8

Combined Operations

Order of operations rules apply to negative numbers also. This means that you must follow the PEMDAS or BODMAS rules.

So to solve $72 \div (-2 + 14) - 11$

First you would complete the work in the parentheses $(-2 + 14) = 12$

Then complete any multiplication and/or division $72 \div 12 = 6$

Then complete any addition and/or subtraction $6 - 11 = -5$

$\therefore 72 \div (-2 + 14) - 11 = -5$

Complete the following Combined Operations Questions

Questions that have the word calculator will require a calculator to complete the question. All answers are to be to 2 decimal places.

a) $10 - 11 + 6 - 7$

b) $-21 \div 7 \times 10 \div -5$

c) $(-48 \div -4) \div (-1 + 3)$

d) $24 \div -3 - 5 \times -2$

e) $81 \div (9 - 18) + (16 - 25)$

f) $(-33 \div 11 + 28 \div -4) \times -10$

g) $26 - 17 \div -8 \times -4 + 3$ (calculator)

h) $81 \div -9 \times -2 + (14 - 17)$

i) $33 \div -11 + 17 \times -6 \div -2$

j) $-17 + 52 \times -2 \div 3 - 21 \div 4$ (calculator)

Note: All these rules for positive and negative numbers apply the same way for fractions and decimals.

Multiplication Tables

To make calculations really easy, learn your multiplications tables. Here is a set of multiplication tables from 1 x 1 to 12 x 12 to help you if you need it.

1 x 1 = 1	2 x 1 = 2	3 x 1 = 3	4 x 1 = 4
1 x 2 = 2	2 x 2 = 4	3 x 2 = 6	4 x 2 = 8
1 x 3 = 3	2 x 3 = 6	3 x 3 = 9	4 x 3 = 12
1 x 4 = 4	2 x 4 = 8	3 x 4 = 12	4 x 4 = 16
1 x 5 = 5	2 x 5 = 10	3 x 5 = 15	4 x 5 = 20
1 x 6 = 6	2 x 6 = 12	3 x 6 = 18	4 x 6 = 24
1 x 7 = 7	2 x 7 = 14	3 x 7 = 21	4 x 7 = 28
1 x 8 = 8	2 x 8 = 16	3 x 8 = 24	4 x 8 = 32
1 x 9 = 9	2 x 9 = 18	3 x 9 = 27	4 x 9 = 36
1 x 10 = 10	2 x 10 = 20	3 x 10 = 30	4 x 10 = 40
1 x 11 = 11	2 x 11 = 22	3 x 11 = 33	4 x 11 = 44
1 x 12 = 12	2 x 12 = 24	3 x 12 = 36	4 x 12 = 48

5 x 1 = 5	6 x 1 = 6	7 x 1 = 7	8 x 1 = 8
5 x 2 = 10	6 x 2 = 12	7 x 2 = 14	8 x 2 = 16
5 x 3 = 15	6 x 3 = 18	7 x 3 = 21	8 x 3 = 24
5 x 4 = 20	6 x 4 = 24	7 x 4 = 28	8 x 4 = 32
5 x 5 = 25	6 x 5 = 30	7 x 5 = 35	8 x 5 = 40
5 x 6 = 30	6 x 6 = 36	7 x 6 = 42	8 x 6 = 48
5 x 7 = 35	6 x 7 = 42	7 x 7 = 49	8 x 7 = 56
5 x 8 = 40	6 x 8 = 48	7 x 8 = 56	8 x 8 = 64
5 x 9 = 45	6 x 9 = 54	7 x 9 = 63	8 x 9 = 72
5 x 10 = 50	6 x 10 = 60	7 x 10 = 70	8 x 10 = 80
5 x 11 = 55	6 x 11 = 66	7 x 11 = 77	8 x 11 = 88
5 x 12 = 60	6 x 12 = 72	7 x 12 = 84	8 x 12 = 96

9 x 1 = 9	10 x 1 = 10	11 x 1 = 11	12 x 1 = 12
9 x 2 = 18	10 x 2 = 20	11 x 2 = 22	12 x 2 = 24
9 x 3 = 27	10 x 3 = 30	11 x 3 = 33	12 x 3 = 36
9 x 4 = 35	10 x 4 = 40	11 x 4 = 44	12 x 4 = 48
9 x 5 = 45	10 x 5 = 50	11 x 5 = 55	12 x 5 = 60
9 x 6 = 54	10 x 6 = 60	11 x 6 = 66	12 x 6 = 72
9 x 7 = 63	10 x 7 = 70	11 x 7 = 77	12 x 7 = 84
9 x 8 = 72	10 x 8 = 80	11 x 8 = 88	12 x 8 = 96
9 x 9 = 81	10 x 9 = 90	11 x 9 = 99	12 x 9 = 108
9 x 10 = 90	10 x 10 = 100	11 x 10 =110	12 x 10 = 120
9 x 11 = 99	10 x 11 = 110	11 x 11 = 121	12 x 11 = 132
9 x 12 = 108	10 x 12 = 120	11 x 12 = 132	12 x 12 = 144

Answers

Indicating Positive or Negative

a) P b) N c) N d) P e) N f) N g) P h) N i) P j) N

Indicate the directed number

a) –$550 b) –15 m c) +7 d) +$400 e) –3° f) +$50 g) +5 h) +$1000 i) –3 j) +3

Inserting the correct symbol between the numbers

a) > b) > c) < d) < e) > f) > g) > h) < i) > j) <

Arrange from smallest to largest

a) –12, –4, –3, –1, 6 b) –11, –4, –2, –1, 1 c) –90, –78, –33, 2, 49 d) –9, –8, –5, –3, 2 e) –89, –12, –7, 0, 76

Arrange from largest to smallest

a) 10, 7, 1, –8, –12 b) –2, –4, –6, –7, –9 c) –2, –5, –9, –11, –12 d) 60, 6, 0, –55, –81 e) 98, 9, –8, –48, –99

Adding negative numbers

a) +17 b) +2 c) +2 d) –2 e) 0 f) –34 g) +3 h) +24 i) +14 j) –18

Subtracting negative numbers

a) –1 b) +2 c) –3 d) 20 e) –9 f) 80 g) –2 h) 0 i) –90 j) 15

Fill in the blanks

a) 12 + 3 which equals 15   b) –5 + 8 which equals 3   c) –2 –7 which equals –9   d) 8 + 40 which equals 48   e) –87 + 32 which equals –55   f) 4 + 5 + 6 which equals 15   g) –8 + 7 – 3 which equals –4   h) –6 –3 –2 which equals –11   i) 2 –8 +3 which equals –3   j) 56 –11 +4 which equals 49

Mixed addition and subtraction questions

a) 11   b) –5   c) 6   d) –26   e) –16   f) 6   g) –9   h) 3   i) –1   j) 7

Multiplying negative numbers

a) –84   b) –15   c) 30   d) 25   e) –16   f) 36   g) –60   h) 64   i) –50   j) –54

Dividing negative numbers

a) 4   b) –3   c) 4   d) –7   e) –6   f) 7   g) –16   h) 9   i) –4   j) 6

Combined operations questions

a) –2   b) 6   c) 6   d) 2   e) –18   f) 100   g) 20.50   h) 15   i) 48   j) –56.92

Glossary of Useful Terms

Integer refers to all positive and negative whole numbers including zero.

Real Numbers refers to all positive and negative numbers including fractions, decimals, etc. It includes all numbers that can be placed on a number line in both directions to infinity.

Natural Numbers refers to all positive integers.

Sum refers to addition. The sum of two numbers is the answer of one number **plus** another number. E.g. the sum of 2 and 6 is 8, (2 + 6 = 8).

Difference refers to subtraction. The difference between two numbers is the answer of one number **minus** another number. E.g. the difference between 6 and 2 is 4, (6 − 2 = 4).

Product refers to multiplication. The product of two numbers is the answer of one number **times** another number. E.g. the product of 2 and 6 is 12, (2 × 6 = 12).

Quotient refers to division. The quotient is the answer of one number being **divided** by another number. E.g. the quotient of 6 and 2 is 3 (6 ÷ 2 = 3).

$\therefore$ Abbreviation for therefore

$\because$ Abbreviation for because

Thank you so much for purchasing this book. I hope it is as useful to you as it has been for my students. If it is, I would appreciate a review on the Amazon website if you would spare a few minutes.

Thank you once again.

Robert

Check out all the books in the Easy Steps Math series at www.amazon.com

Fractions
Decimals
Percentages
Ratios
Negative Numbers
Algebra
Master Collection 1 – Fractions, Decimals and Percentages
Master Collection 2 – Fractions, Decimals and Ratios
Master Collection 3 – Fractions, Percentages and Ratios
Master Collection 4 – Decimals, Percentages and Ratios

Made in the USA
Lexington, KY
08 August 2019